The Bee Keeper's Guide

Containing concise practical directi~
management of bees, upon ·'

J. H. Payne

Alpha Editions

This edition published in 2024

ISBN : 9789367246979

Design and Setting By
Alpha Editions
www.alphaedis.com
Email - info@alphaedis.com

As per information held with us this book is in Public Domain.
This book is a reproduction of an important historical work. Alpha Editions uses the best technology to reproduce historical work in the same manner it was first published to preserve its original nature. Any marks or number seen are left intentionally to preserve its true form.

Contents

PREFACE TO THE FIRST EDITION...- 1 -

PREFACE TO THE SECOND EDITION..- 3 -

CHAPTER I. ....- 5 -

CHAPTER II. ....- 9 -

CHAPTER III. ....- 11 -

CHAPTER IV. ....- 14 -

CHAPTER V. ....- 18 -

CHAPTER VI. ....- 21 -

CHAPTER VII. ....- 23 -

CHAPTER VIII. ....- 25 -

CHAPTER IX. ....- 28 -

CHAPTER X. ....- 30 -

CHAPTER XI. ....- 31 -

CHAPTER XII. ....- 34 -

CHAPTER XIII. ....- 37 -

CHAPTER XIV. ....- 38 -

CHAPTER XV...- 40 -

CHAPTER XVI. ....- 41 -

CHAPTER XVII. ....- 43 -

CHAPTER XVIII. ....- 46 -

PREFACE
TO THE FIRST EDITION.

HAVING written the "Cottager's Guide for the Management of his Bees, upon the Depriving System," which has been printed under the direction of the Suffolk and Norfolk Apiarian Society, for gratuitous distribution amongst the Cottagers, I am induced, at the particular request of several Apiarian friends, to enlarge the above little work, and to give in addition a description of Nutt's newly invented Hive, and other practical remarks in Bee-knowledge, resulting from nearly forty years close observation.

Should this little work be the means of inducing any person to promote the culture of Bees amongst the Cottagers in his immediate neighbourhood, upon the Depriving System, I shall be amply repaid for the trouble it may have occasioned me; and the hope that such will be the result, must be my apology for adding to the number of books (perhaps already too numerous) upon this subject.

Reaumur in a letter to M. de la Bourdonaye, in 1757, says, "The preservation and also the increase of Bees is an object of such interest to Britanny, that the peasantry cannot be too much encouraged to turn their attention to it." Surely this is equally applicable to our own country at the present time, when the condition of the labouring poor calls so loudly for relief.

I have little hesitation in saying, that Cottagers who are able to keep from four to six Hives of Bees, may make from four to eight pounds, yearly profit, after paying all expenses upon them. I paid last year to one Cottager seven pounds, fifteen shillings, and to another five pounds and one shilling for Bees and Honey.

The following anecdote has so much the appearance of truth in it, and is so well suited to my present purpose, that I cannot refrain from giving it.

A good old French Bishop in paying his annual visit to his Clergy, was very much afflicted by the representations they made of their extreme poverty, which, indeed, the appearance of their houses and families corroborated. Whilst he was deploring the state of things which had reduced them to this sad condition, he arrived at the house of a Curate, who, living amongst a poorer set of parishioners than any he had yet visited, would, he feared, be in still more woful plight than the others; contrary however to his expectations, he found appearances very much improved, every thing about the house wore the aspect of comfort and plenty. The good Bishop was amazed. "How is this, my friend?" said he, "you are the first man that I have met with a cheerful face and a plentiful board. Have you any income independent of your Curé?"

"Yes, Sir," said the Clergyman, "I have; my family would starve on the pittance I receive from the poor people that I instruct, come with me into the garden and I will show you the *Stock* that yields me an excellent interest."

On going to the garden he showed the Bishop a large range of Bee-hives.

"There is the Bank," he continued, "from which I draw my annual dividend.—It never stops payment."

Ever after that memorable visit, when any of his Clergy complained to the Bishop of poverty, he would say to them, "*Keep Bees! Keep Bees!*"

In the words of an Apiarian friend, I solicit information from every one who may have it in his power to transmit it to me, and on the other hand, I profess my perfect readiness to impart whatever knowledge I may possess in the management of an Apiary, to any person who will favour me with the application; my aim is general utility, and the establishment of a national advantage.

PREFACE
TO THE SECOND EDITION.

SINCE the first appearance of this little Treatise I am most happy in being able to state that Apiarian Science has in this neighbourhood and in the adjoining Counties, made very considerable advances, that the ridiculous notions, and foolish prejudices entertained respecting Bees, are fast wearing away—that the *Cottagers* are generally managing them upon the depriving system, making them a real source of profit and of comfort, and that a number of influential persons are making themselves acquainted with the practical management of Bees, upon the simplest and most profitable methods, for the sole purpose of setting an example, and for qualifying themselves to give instructions in the management of them to their poor neighbours. Nor is this spirit of well directed benevolence confined to these Counties only, for at Oxford a Society is just formed to promote an improved and more extensive system of Bee management among the Cottagers by the diffusion of information on the subject, and the *loan* of hives *not the gift*, their cost to be repaid from the produce, and also to promote a more extensive and scientific knowledge concerning the Natural History and cultivation of Bees among the higher classes; the society I find is flourishing, a piece of ground has been taken and laid out as an experimental Bee-garden, there is already a very considerable number of stocks of Bees placed in it in common straw, and experimental hives. Subscribers pay half a guinea a year, and non-Subscribers a shilling each visit This is an example worthy of imitation.

I am just favoured with a letter from a Gentleman who has recently visited the above establishment giving me a certain and simple method of Autumnal union of weak stocks, which he there witnessed, but it comes too late to be embodied in this treatise the whole of that part being already in the press, still as I consider the union of weak stocks important, and Gelieu's method which I have given too troublesome for most persons, I will venture to give it in this place. "The process" he says, "is merely *fumigating the Bees* for which they have invented a tube much more simple than Nutt's which they insert into the mouth of the hive; under the hive is previously pushed either an empty one reversed, or a shallow box with a wide rim, this receives the stupified Bees; cut out the combs and remove all the remaining Bern from them that none be lost. Now take a little sugared ale and sprinkle it over them just as they are recovering, place upon them the hive to which they are to be united, this hive requires no sprinkling nor any fumigation, the Bees in the latter are soon attracted by we ale and go down into the hollow containing the fumigated ones licking them over, the whole are mixed and go up without the least disturbance, it is unnecessary to take any trouble about the Queen," he adds "I was assured that not a Bee would be lost" and he further says "upon my return home I tested it with entire success on some of my neighbour's Bees, it was the work of but a few minutes, and not the smallest danger. I left the hive placed upon the other all night, and the next morning every Bee had left the bottom one, more perfect quiet I never saw. I think there were nearly two quarts of Bees fumigated."

Puff-ball is generally recommended being the safest, mild tobacco answers very well, great caution, however, is required in its use, or the Bees may be killed. Common fumigating bellows, or even a tobacco-pipe may be used for this operation: After this discovery it will be absurd not to unite weak stocks, or to destroy a single Bee on taking up an old hive.

I have always considered the keeping of Bees and the advantages arising from them to be the undoubted privilege of the Cottagers and theirs alone, other persons may keep a few for amusement, or to endeavour to learn something of their natural history, but all should in my opinion be made subservient to the Cottagers' benefit.

The present season has been a most unpropitious one, especially in this neighbourhood, perhaps the most so that I remember, but I am disposed to think that this is not a favourable district for the collection of honey. I have frequently compared the produce of my own apiary with that of others at a distance, and this year especially, with that of a friend upon the Essex bank of the Stour, which I assisted in forming and have frequently visited, the quantity of honey obtained from this is small, but it is five times as much as that of any one in this neighbourhood consisting of the same number of stocks.

I cannot close this preface without acknowledging the very flattering manner in which my little treatise has been mentioned in various publications, and my thanks to the many correspondents it has obtained for me. The addition which I have been enabled to make to it, will I trust be acceptable and useful.

I still solicit information from any who will give it me, and am ready to impart it to all those who will ask it.

Bury Saint Edmund's,
    Oct, 11th, 1838.

CHAPTER I.

Situation of an Apiary, and directions for placing the Hives.

I HAVE no hesitation in saying, that a South aspect is decidedly preferable to any other situation for an Apiary. I have tried various aspects, but the Bees in the South I have always found to be the healthiest, and to collect the largest quantity of honey. It is very important that the hives be sheltered from the wind by trees or houses, and that they are not placed in the vicinity of ponds or large rivers, for high winds will dash them into the water, where numbers will perish.

It was the opinion of the ancients, that Bees in windy weather carried weights to prevent them from being driven about by it. Virgil says,

"That with light pebbles, like a balanced boat,

Poised through the air on even pinions float."

SOTHERBY'S GEORGICS.

This is now ascertained to be erroneous, and is ascribed by Swammerdam and Reaumur, to preceding observers having mistaken the Mason Bee, for a Hive Bee; the former builds its nest against a wall, with a composition of gravel and its own saliva, and when freighted with the former article, may easily have led a careless observer into the erroneous opinion alluded to. The Abbe della Rocca appears to have fallen into, and perpetuated, the same error.

Though large ponds are very injurious, a small stream is beneficial to them, which if they are not supplied with, water must be given them, for it is absolutely necessary, and enters, as much as honey and farina, into the composition with which they nourish the brood. The plan that I have for many years adopted, is to fill an unglazed earthen pan, eighteen inches by eight, four inches deep, and square at the sides, with water, upon the surface of which floats a very thin deal board perforated with holes: in Spring and Summer, the Bees may be seen coming in great numbers to drink, or rather to carry water into their hives to mix with the farina they collect so abundantly at this season of the year for food for their young. In my opinion, Dr. Bevan says very justly, that "the Apiary should be near the residence of the proprietor, as well for the purpose of rendering the Bees tractable, and well acquainted with the family, as for affording a good view of their general proceedings."

I am a decided enemy to Bee houses of all kinds for they are the means of causing the ruin of a great number of hives, by affording a home to their worst enemies, viz. mice, moths, spiders, earwigs, and various other insects, thousands die from imprisonment, and many hives are destroyed by humidity. The method of placing several hives upon the same bench is also very injurious, it very much facilitates pilfering, and renders it impossible to operate upon one hive, without disturbing the others.

The hives should be placed upon separate boards, supported by single pedestals four or five inches in diameter, firmly placed in the ground, and standing about fifteen inches from the surface, (see fig. 1. plate 1.); upon the top of this post should be nailed firmly a board nine inches square, upon which should be placed the board the hive stands upon, but not nailed, the double boards will be found very convenient for weighing or removing the hives, without disturbing the Bees.

On no account use clay or mortar as is usually done to secure the hive to the board, the Bees of themselves will do it more effectually;[1] clay or mortar tends very much to decay the hives, and to harbour moths and other insects; each hive should be covered with a large milkpan, and be well painted every year, for hives managed upon the depriving system, are expected to stand from fifteen to twenty years.

[1] This fact, though it has been denied by those who profess to have had much experience in the management of Bees, is known to every novice in Apiarian science, for he does not suffer much time to pass, after having purchased a swarm of Bees, without endeavouring to ascertain how much honey they have collected, and finds the difficulty of separating the hive from the board upon which it was placed.

The hives should be placed about three feet apart from each other, and in a right line, but should the number be too great to allow of this arrangement, and render two rows necessary, they must not be less that fifteen feet asunder, and those in the front row intersecting the line formed by the hinder one.

The boards on which the hives are placed, should be cleaned about four times in the year, January, March, April and November, much time and trouble will be saved the Bees thereby.

Plants which rise in height equal to or exceeding the entrance of the hives, should not be suffered to grow in their immediate vicinity, and every facility should be removed by which the enemies of the Bees can ascend into the hives.

Still, however, a few shrubs or standard roses of four or five feet may with advantage be placed eight or ten paces in front of the hives, for the Bees to alight upon in their return home when heavily laden with honey and pollen—it saves their falling to the ground from the weight of their load, which they frequently do, and in unfavourable weather to rise no more—it was seeing them rest in this manner that gave rise to the following lines:—

Rest on that Rose's leaf awhile, thou little Busy Bee,

Thou hast winged thy way with thousands, the wand'ring, the free,

Unwearied with thy ceaseless toil in search for future store,

Thou'rt comeback to unlade thy sweets, then sally forth for more.

Thou'st been among the flow'rs of gold, their kiss is on thee yet,

And o'er thy richly powdered wings how many hues are met,

That tell of revelling at the founts of nectar's luscious tide,

Of honey-dews that rest upon each petal's glossy side.

Where hast thou been since the bright morn first saw thee on thy way

'Mong scented brier and glittering heath that woo'd thy lingering stay?

Hast thou no voice to tell us of the far off verdant scenes,

Of the rich limes thou lov'st so well and of the fresh'ning steams.

Away! away! once more thou'rt up and ev'r the leaf be still'd.

To its soft rest from the trembling that thy light form has thrill'd,

Thou'lt be again among thy loves, the fragrant, the bright,

All jealous of their hidden sweets, in murmuring delight.

I have always found the advantage of planting, in the vicinity of my hives, a large quantity of the common kinds of crocus, single blue hipatica, heleborus niger, and tussilago petasites, all of which flower very early and are rich in honey and farina: salvia nemorosa, (of Dr. Smith) which flowers very early in June and lasts all the summer, is in an extraordinary manner sought after by the Bees, and when room is not an object, twenty or thirty square yards of it may be grown with advantage, origanum humile, origanum rubescens, (of Haworth) and mignonette may also be grown; cultivation beyond this, exclusively for Bees, I believe answers very little purpose.

Doctor Bevan says: "To those who, residing in towns, may consider it indispensable to the success of an Apiary, that it should be in the immediate vicinity of good pasturage, and be thereby deterred from benefiting and amusing themselves by keeping Bees; it may be satisfactory to learn that the Apiary of the celebrated Bonner was situated in a garret, in the centre of Glasgow, where it flourished for several years, and furnished him with the means of making many interesting and valuable observations which he gave to the world about thirty years ago."

My own experience also proves the truth of the above statement, residing myself for four years in the centre of a large town, in a house without a garden, I kept two stocks of Bees in my study, in glass, and four or five others in the improved cottage hive upon the roof of my house, and I am not aware that they have ever done better, or afforded me a larger quantity of honey in any other situation.

CHAPTER II.

Directions for purchasing Bees.

THE best time to establish an Apiary is from the middle of February to the middle of March, the stocks will have passed in safety through the winter, the combs are then empty of brood, light of honey, and the removal safe and easy. Stocks should be selected by a competent judge, as the weight alone cannot be relied on, a swarm of the preceding year should be selected, and one that contains not less than twelve pounds of honey; there are few commodities in which a person can be so easily deceived as in a hive of Bees. I would therefore recommend the young Apiarian to take the opinion of some experienced person before he makes his purchase, a hive of the preceding year can only be known by a close inspection of the combs, which but few persons have courage to engage in; if the hive is not of the preceding year its weight is no criterion of its value, for an old hive always contains a large quantity of the pollen or dust of flowers which the Bees carry home on their legs, especially in the Spring and Autumn, it is an essential ingredient in the food with which they nourish their young, but good for nothing else, indeed the Bees will die of hunger upon the combs that are filled with it:—"Yet," says Gelieu, "they lay up useless hoards of it, which they go on augmenting every year, and this is the only point on which they can be accused of a want of that prudence and foresight so admirable in every other respect."

The Bees appear to be aware of the perishable nature of this substance, for they never fill a cell entirely with it, but leave room for a small quantity of honey in each cell containing pollen, before it is sealed up, by this means the air is most effectually excluded, and the pollen preserved for a considerable time; should, however, the Bees be compelled to consume the honey from those cells containing pollen, before they can make use of it for their young, it moulds and becomes of no value, and causes them great labour to remove it. For when in this state, they have no means of displacing it but by eating away the cells in which it is contained, and conveying it out of their hives in small pieces, about the size of peas, hard and mouldy. I have seen the entrance of old hives in the month of April almost filled up with the pellets of mouldy farina. The process is tedious, takes up much time, and the ravages made by it upon the combs appear irreparable; still in a short space of time, if the weather be favourable, the combs are repaired, as if no injury had befallen them, and filled with honey or brood. It is a very heavy substance, so that if weight be the only criterion, farina will be purchased instead of honey, therefore in the purchase of old stocks it will be necessary they should weigh eight pounds more than swarms of the preceding year; in the purchase of swarms less experience is necessary, and by attending to the following rules the young Apiarian will not be imposed upon.

1st. That the swarm be purchased before the 14th of June, the longer before that time the better.

2ndly. That it does not weigh less than three pounds and a half. I have known some swarms to weigh six pounds, but this is of rare occurrence.

It is very important to observe, that when a swarm of Bees is purchased it must be removed to the place in which it is to remain, upon the evening of the day it swarmed, for should the removal be delayed even till the next day, the combs will in all probability be broken and the stock destroyed.

I should recommend the purchaser to send his own hive to the person of whom he intends to buy a swarm, and to desire him not to put any sticks across the interior of the hive, as is the usual custom, for they cause much trouble to the Bees in forming their combs, and render their extraction almost impossible. The prosperity of the hive will much (perhaps entirely) depend upon its being finally placed upon the evening of the day it swarmed.

CHAPTER III.

Materials of which Hives should be made, and the improved Cottage Hive described.

MUCH has been said with respect to the materials of which Hives should be made, and experience has long determined, that straw and wood are the best. Mr. Huish, to whom I am indebted for some useful information in Apiarian science, says, "Of all the materials which have been selected for the formation of a Hive, I conceive no one to be more eligible than straw." Gelieu, to whom experience as an Apiarian I am disposed to pay the greatest respect, and whose work containing Practical Directions for the Management and Preservation of Hives,[2] I would recommend to every keeper of Bees, says, (when speaking of wood and straw as materials proper for Hives) "experience has shown me, that it is a matter of indifference which are employed; except as to price, according as either material may be more or less abundant in different parts of the country." I have for nine years possessed a Nutt's hive, which is made of wood, (and which I shall have occasion hereafter to mention,) without being able to discover any difference in the health and activity of the Bees; but the facility and economy in the construction of straw hives, must always be a recommendation, as it is in every article connected with rural economy.

[2] The 'Bee Preserver,' by Jonas De Gelieu, recently translated from the French.

It has always been my practice to paint my Hives, both wood and straw, at least once in the year, and I would strongly recommend all persons to do the same. April I think is the best time, and if done after six o'clock in the evening, not the least inconvenience will arise either to the painter or to the Bees.

Having decided upon the materials of which Hives should be made, their form is next to be considered; for a straw hive, I would recommend the following size, nine inches deep and twelve in diameter, straight at the sides and flat at the top, in shape like a half bushel measure, a hole should be made in the top of four inches, and a piece of straw large enough to cover it must be fastened on with skewers, (see fig. 3, plate 1.) not to fit in but to cover *over*, the diameter of the piece of straw being at least two inches more than that of the opening at the top of the hive, it will be much easier taken off, and the combs of swarms of a few weeks standing will not be injured by its removal, which in favourable seasons should always have a glass or small hive put upon them, the early ones especially; an entrance of two inches by one, must be cut in the bottom of the hive, to which I affix a piece of copper of about six inches by three, having a grove, to admit two sliding copper plates, one perforated and the other having a hole large enough to allow but one Bee to come out at a time, (see fig. 10, plate 2.)[3]

[3] Instead of two sliding plates, I would recommend one only as given by Mr. Taylor, in his "Bee-Keeper's Manual," lately-published, for as they cannot both be used at the same time, the one out of use is frequently misplaced. Zinc answers the purpose equally with copper, and is but half its price.

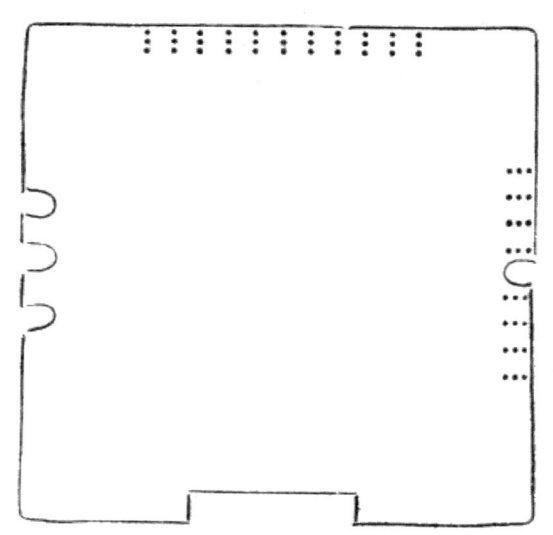

I have found great advantage arising from this little apparatus. The finely perforated slider is used to confine the Bees to their hive when snow lies upon the ground, the reflection of which, when the sun shines upon it, never fails to induce them to leave their hives, and falling upon it they perish, for a Bee becomes torpid at a temperature of 32°. The slider with one hole only, is useful both in Spring and Autumn, preventing either robbers or wasps from entering the hives, for three or four Bees will, with the help of this slider, guard the entrance more effectually that ten times the number without it.

Although I have recommended Bees to be confined in their hives so long as snow remains upon the ground, it would, however, be very prejudicial to them if carried on beyond that time, for I never saw Bees healthy and strong after being shut up through the winter.

Gelieu says, "Bees have no real disease, dysentery, about which so much noise has been made, and for which so many remedies have been prescribed, never attacks the Bees of a well-stocked hive, that is left open at all seasons, but those only that are too long and too closely confined. They are always in good health as long as they are at liberty, when they are warm enough and have plenty of food. All their pretended diseases are the result of cold, hunger, or the infection produced by a too close and long confinement during winter."

CHAPTER IV.

Method of placing the small Hive,[4] Box, or Glass upon the Improved Cottage Hive, by which means fine Honey may be obtained, without destroying the Bees.

[4] To avoid repetition, we shall in future use the term, "Box," to express any receptacle employed to obtain honey on the improved system, whether it be in wood, glass, straw, or any other material.

AT the end of April, or very early in the month of May, take the moveable piece of straw, from the top of the Improved Cottage Hive, (fig. 3.) and place it upon the adapter, (fig. 5,) then put the Box or small Hive (fig. 7, and 4) upon this adapter and cover the whole with a milk-pan, to defend them from wet. A glass may be used instead of the small Hive or Box, with equal success, providing it be covered with something that will effectually exclude light; a cover of straw, is perhaps, preferable to any other.

When the Bees are beginning to work in a glass, a cold night generally obliges them to forsake their newly made combs, sends them down into the hive, and compels them to discontinue their labours which are seldom resumed till the middle of the next day; to prevent this delay I would recommend the space between the glass and its cover to be filled with fine tow or wool, the temperature of the glass being thereby kept up, and the Bees enabled to carry on their labours without interruption.

Experience has proved that the milk-pan is the best of all protections for a hive, provided it be six inches in diameter larger than the hive itself.

When the Box is filled with honey and the combs partially sealed, or when the Bees are seen to cluster at the mouth of the Hive at nine or ten o'clock in the morning, let no time be lost in lifting up the Box, and placing between it and the Stock-hive another Box with a hole in the top; the adapter (fig. 5) will be found very useful in this operation. It is necessary to use this precaution at all times, but more especially in a rainy season as a greater disposition amongst the Bees to swarm then prevails. "Dry weather makes plenty of honey, and moist, of swarms."[5] However incorrect this position may at first sight appear, the attentive observer will quickly become convinced of its truth.

[5] Purchas on Bees.

Since the publication of the first edition of this little Treatise, many persons have said to me, "their Bees would swarm, although the small hive had been placed on as directed above, and sometimes after they had commenced working in it," the reason for which in my opinion is, that the second small hive was not supplied soon enough, for the like has never in a single instance occurred with my own Bees. I have not had a swarm these twenty years from any of the hives worked upon the Depriving System, occasionally I have compelled a hive to swarm, to fill up a vacancy in my number, where the Queen has died, or some other accident destroyed the stock.

The population of a hive increases rapidly in April and May, and consequently the internal temperature rises in proportion, a very high temperature causes swarming, (Mr. Nutt says 130°) although the Bees may have abundance of room—I have frequently seen a glass lamp that has no opening at the top, placed upon a hive, and the result has been that the Bees swarmed before they had filled it.—If both *room and ventilation* are carefully attended to *swarming may be prevented altogether*, and that the one may be as completely under the control of the proprietor as the other, I would recommend Mr. Taylor's Ventilator, which I believe, to be a perfect one, for when properly arranged it will reduce the temperature of a hive at the swarming season, from ten to twenty degrees in a few minutes—I would recommend its insertion in the top of the small hive, box, or glass, before it is placed upon the larger one.

"The Ventilator I use, says Mr. Taylor, (and I have made them of various forms) consists of double tubes, both resting on a flauch in the holes prepared for them, the outer tube is of one inch diameter and six inches long, with six half inch holes dispersed over it.

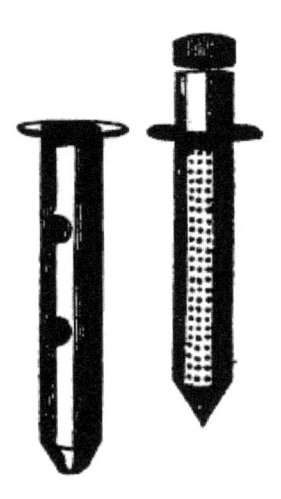

It is soon fixed down in its place by the Bees, and so must remain, the inner tube is of perforated zinc, with a[6] tin projecting top as a handle, and a cap to put on or off this as required. The Bees will stop up the inner tube where they can get at it, when it may be turned round a little to present a new surface. When wholly stopped, it may be withdrawn from its place, and a clean tube substituted. This may be done without the least danger to the operator, but it should be inserted carefully, to avoid crushing any Bees that may have crept within the outer tube, an exit to these is afforded by the hole at the bottom. The substance with which Bees glue up all crevices and attach their combs is called Propolis—a resinous exudation from certain trees, of a fragrant smell, and removable by the aid of hot water.

[6] In adopting Mr. Taylor's Ventilator to the small hive, the inner tube must be made without "the projecting top as a handle," and the cap made even with the flauch.

In order occasionally to know the temperature of any of the boxes, a thermometer made to fit the ventilator may be inserted in it. This is chiefly useful as a matter of precaution towards the swarming season.

Some have thought it necessary to cut holes in the floor-board, or have placed drawers underneath, to aid the ventilation, but I have always found the mouth of the hive sufficient for all purposes."

All operations except joining swarms or stocks, should be performed upon a fine day, about noon, they may then be done with much less annoyance to the Bees, as well as with less chance of danger to the operator.

I have for some years past performed almost all the operations required in this system without the defence, even of gloves, but I would not recommend any person to attempt it, until he has had several years experience in the management of Bees.

The being perfectly defended in every part against their stings, gives that coolness and confidence to the operator, upon which the happy accomplishment of his intentions so much depends. I cannot too strongly urge, that coolness and confidence on the part of the operator are essential qualifications, for anything approaching to hurry irritates them beyond measure; indeed whilst engaged with them the hand ought never to be hastily removed from one position to another. Dr. Bevan says, "quietness is the surest protection against being stung."

The best defence is a mask of wire, very similar to, but much finer than a fencing mask, with a rim of tin made to fit the head, to which a silk handkerchief is attached, a pair of thick worsted gloves, and stockings or gaiters of the same material; stout leather gloves are as good protection as those of worsted, but leather, from the closeness of its texture, will not allow the Bees to withdraw their stings from it and the consequence is, that many perish.

It is recommended to persons during their operations on Bees, to carefully avoid breathing upon, them, as nothing is more offensive, or more irritating to them than the human breath; this however, is partially obviated by closing the mouth, and suffering the breath to pass gently through the nose, by which means a full current is not allowed to fall upon them.

CHAPTER V.

Proper time for taking away the Box and how to expel the Bees from it.

WHEN the Box is filled with honey and the combs all sealed up, (which will generally be done about the middle of June) it may be taken off, or it may remain till the one placed beneath it is also sealed up, which in all probability will be completed by the first week in August.

Upon the very strong and populous hives, it is necessary in some seasons to place even a third, which must be removed with great caution, for at this time of the year every stock should contain at least twenty pounds of honey; should however the stock have that quantity, it may be safely removed and placed upon a weaker one, for the combs not being all sealed, the honey therefore is not saleable. The Box taken off, must be lifted very gently at noon, upon a fine day, and carried forty or fifty yards from the hive; place it upon a board or table, raising it a little that room may be given to the Bees to make their escape, which they will do in a very short time.

Much difficulty appears to have arisen with some persons in getting the Bees to leave the Box when taken off, but in this as in all other operations with Bees, gentleness is very important, indeed it is the only means of accomplishing the end desired, and as I have before said, "the Box must be lifted *very gently*," and placed about six inches from the ground, or table, upon bricks, flower pots, or something of the kind. Shaking, beating, or burning paper under it, as is sometimes done, have all a contrary effect upon the Bees, they are alarmed by it, and will not leave the box perhaps for days, when these means have been resorted to.

The box being thus placed, a loud humming noise is first heard, and the Bees are then seen to leave it within five or six minutes, (all except a few stragglers), but should the Queen be in the Box, (which very rarely happens) quite a different appearance presents itself, no noise will be heard, or a Bee scarcely seen to leave it, but the hive from which it has been taken will in a very short time appear in great confusion. Whenever this occurs, the Box must be returned immediately, and taken off again the next day.

When a hive or glass of honey is taken, it ought not to be left till the Bees are all out of it, for it is very likely to be attacked by robbers, thus a great part of it will be carried away in a short time, and what is left rendered unfit for sale, on account of the cells being opened, from which the honey will drain out upon the position of the hive being changed.

Robbers may be known by their desire to enter the hive or glass, the Bees belonging to it, being separated from their Queen, fly home immediately upon leaving it.

In taking off a box of honey it will be found convenient to pass a very thin knife, or fine wire, between the hives or boxes intended to be separated; if that precaution be not taken, a piece of comb frequently projects from the top of the one left, or the bottom of that taken, which causes much trouble to the operator: two adapters (fig. 5.) placed between the boxes will be found very convenient, for the knife or wire will only have to be passed between them, and the danger of breaking the combs will be obviated—they should be made of mahogany, for it will allow of being worked very thin, without the risk of warping.

To expel the bees from the box or hive when taken off, Gelieu says, "Take a hive or box of the same size, place it over the full one that is turned upside down, bind them round with a napkin, to intercept all passage to the bees, and force them to ascend into the empty box, by tapping gently on the full one. They soon go up into the empty box, and when they are all housed, replace them on the parent hive, whence they were withdrawn; and if the season is favourable and the honey abundant, they soon set to work again."

Honey taken by this method is acknowledged to be very superior in quality to that obtained by the usual barbarous and unprofitable manner of burning the Bees, which arises from the combs in which it is deposited being new and perfectly white, the early period at which it is collected, and from its being unmixed with honey gathered later in the season as well as from the Queen very rarely ascending through the opening at the top of the improved cottage hive, that neither brood nor farina are found amongst it.

This honey sells readily at two shillings a pound, whilst that obtained by burning the Bees, is scarcely saleable at eightpence.

It is usual to obtain from every good stock twenty or perhaps thirty pounds of honey annually. I once obtained forty-five pounds, leaving the stock rich in honey.

It is frequently asked what becomes of the Bees managed on this system, if they are never suffered to swarm nor are destroyed;—the hives will never contain them? To which I would reply, that it is well known to those who are conversant in the care of Bees, that their numbers decrease greatly in Autumn, not only by the destruction of the drones, but also by the unavoidable deaths of many of the working Bees, owing to the thousand accidents they meet with in the fields, and to age;[7] a much less space therefore is wanted for them in the winter than was necessary in the summer.

[7] Mr. Purchas in his "Theatre of Political Flying Insects," published in 1657—says, "it is manifest that the Honey Bees are but yearly creatures," and when giving the sentiments of Aristotle, Pliny, Columella, Cardanus, and others, he says, "the truth is, notwithstanding these men's opinions, that Bees live but a year and a quarter at most, for those Bees that are seen in May, lusty, full, brown, smooth, well winged, will by the end of July following, begin to wither, become less*e*, look gr*ay*, and have their wings t*o*ttered, and torn, and be*e* all dead before the end of August."

CHAPTER VI.

Method to be pursued in case a Swarm should leave the Hive, after having commenced working in the Box.

THIS is a circumstance of very rare occurrence, and more especially when the directions given in the former chapter are strictly complied with; however, should it happen, let the swarm be hived in the usual manner into the improved Cottage Hive, (see fig. 3.) as directed in Chapter XII, when the Bees are settled, take off the moveable piece of straw from the top of the hive, and place upon it the box partially filled with honey and Bees; cover the old hive with the piece of straw belonging to it, and the milkpan, as no further profit (except the cast or second swarm) will be obtained till the next season; should the proprietor be unwilling to increase his number of stocks, the swarm may be returned immediately to its parent hive; the process is very simple, and I have always found it succeed—as soon as the swarm is settled, turn the hive bottom upwards, and if the Queen Bee does not make her appearance in a few seconds, dash the Bees out upon a cloth, or upon a gravel walk,[8] and with a wine glass she may be easily captured, upon this being accomplished, the Bees will immediately return to their parent hive and resume their labours; she may also very easily be taken during the departure of a swarm, for she appears to leave the hive reluctantly, and may be seen running backwards and forwards upon the alighting-board before she takes wing.

[8] The method of performing this operation, consists in lifting the hive gently about a foot, and with a smart and sudden jar returning it to the ground, so that the Bees be completely dislodged from the hive and left upon the cloth, the hive may now be removed to a short distance, and as the Bees are attempting to return to their former habitation the Queen may be easily captured.

A second swarm generally leaves the hive about nine days after the first, but the time may be exactly ascertained by standing quietly beside the hive after sunset, when the Queen may be distinctly heard "to tun' in hir treble voic',"[9] which is a certain indication that a second swarm will leave the hive. Should two or three Queens be heard one after the other, it will be on the following day, if the weather be not *very* unfavourable, (for the second and third swarms appear to have less regard as to the weather than the first.) Should the Queens continue to pipe after the departure of the second swarm, a third will certainly follow in a few days, but if one or two Queens be found dead beneath the hive on the next morning, no more swarms can be expected.

[9] Butler's Feminin Monarchi—Edit. 1634.

That the old Queen accompanies the first swarm is established beyond a doubt; that many Queens are bred in a hive, a number sometimes exceeding thirty in one year, is also ascertained; and that the Bees have the power of producing a Queen from an egg deposited in the combs of the working Bees, by treating it in a different manner to those that are to become workers, has also been satisfactorily proved, all that has been said beyond this, regarding their natural history, must, I believe, be considered principally conjecture.

It is, however, says a modern Author, "not the least interesting part of the study of the Bee, that this apparently insignificant insect has hitherto baffled all the research and ingenuity of man to discover the manner of its propagation; analogy presents no guide to the solution of this secret, and the result of every anatomical experiment has tended rather to mystify the subject, than to conduct us to the road to truth," and Purchas, who I have before quoted says, "God humbles us with ignorance in many things, not only divine but natural and in common use, in the nature of Bees how blind are we, notwithstanding all our observations and labour in the production and continuance of the Queen Bee, in the generation of other Bees, and generally in the forms of all things."

CHAPTER VII.

Method of uniting second and third Swarms.

SECOND and third Swarms, or Casts and Colts, are seldom or ever able to collect a sufficient quantity of honey, to support themselves through the winter, and can only be preserved by much care and expense, and most of them die after all without bringing any profit. It is much better therefore to unite them in the following manner:—when two *Casts* or *Colts* come off upon the same day hive them separately and leave them till an hour and half after sunset, then spread a cloth upon the ground, upon which by a smart and sudden movement shake all the Bees out of one of the hives, and immediately take the other and place it gently over the Bees that are heaped together upon the cloth, and they will instantly ascend into it and join those, which, not having been disturbed, are quiet in their new abode; next morning before sunrise remove this newly united hive to the place in which it is to remain; this doubled population will work with double success and in the most perfect harmony, and generally become a strong stock from which much profit may be derived.

Two Casts or weak Swarms may be joined in the same manner, although one of them may have swarmed some days or even weeks later than the other, taking care however not to make the first one enter the second, but the second the first, a third and a fourth parcel of Bees may be joined to them at different times till the stock becomes strong.

It is almost impossible sufficiently to impress upon the mind of every person who keeps Bees the necessity of having his stocks all strong, for weak stocks are very troublesome, very expensive, and seldom, if ever, afford any profit.

Mr. Taylor says, "the stronger the colony at the outset the better the Bees will work, and the more prosperous it will become. I never knew a weak one do well long, and a little extra expense and trouble at first are amply rewarded by succeeding years of prosperity and ultimate profit;" and again, "thus strength in one year begets it in succeeding ones, and this principle ought to be borne in mind by those who imagine that the deficient population of one season will be made up in the next, and that the loss of Bees in the winter is of secondary consequence, forgetting how influential is their warmth to the earlier and increased productive powers of the Queen, and how important it is in the opening spring to be able to spare from the home duties of the hive a number of collectors, to add to the stores, which would otherwise not keep pace with the cravings of the rising generation."

It is a remarkable fact, that two weak stocks joined, will collect double the quantity of honey, and consume much less than two of the same age and strength kept separately. Stocks must be joined after sunset upon the day that one of them has swarmed, or before sunrise the next morning, and the doubled stock must be placed upon the stand it previously occupied. Great care must be taken not to shake the hive, nor must it be turned up, the combs being new, and tender, will easily break, and the stock by that means be destroyed.

CHAPTER VIII.

Manner of uniting Swarms and old stocks in Autumn.

FOR this very useful information I am indebted to that excellent Apiarian, Gelieu; I have tried it upon some of my own stocks, as well as upon those of my friends, and have found it in every instance fully to answer my expectations. Persons possessing these instructions should not allow a weak stock to remain through the winter.

The operation is performed very easily and without danger: I have frequently accomplished it without any protection whatever, and I will give the method in the words of Gelieu—"When the swarms have not been able to lay up a sufficient provision during the fine weather, I weigh them at the end of the season, and knowing the weight of each empty hive, I can tell exactly the quantity of honey they have in store. If they are three, four, five or six pounds too light, I preserve them and feed them in the manner I am about to detail. When the swarms have only about one-third or one half of the quantity of honey which would suffice to feed them, I might keep them alive by giving them as much more as they require. I have frequently done so, but I have already remarked that this plan costs too much honey, and gives too much trouble: and, therefore, I generally join them into one. For this purpose, I leave the heaviest swarm untouched, and, in the morning of a fine day in September, or the beginning of October, I commence by blowing a few whiffs of tobacco-smoke with my pipe in at the door of the hive of the lightest swarm, to disperse the sentinels; then turning up the hive, and placing it on its top on the ground, I give it a little more smoke, to prevent the Bees from becoming irritated, and to force them to retire within the combs—I proceed to cut out all the combs in succession, beginning with the smallest, sweeping the Bees with a feather off each piece back into the hive; and then I place the combs, one after another, into a large dish beside me, keeping it, at the same time, carefully covered over with a napkin, or small table cloth, to prevent the Bees returning to their combs, or the smell of the honey attracting others that may be flying about. The last comb is the most difficult to come at, being completely covered over with Bees. I detach it, however, in the same way as the others, but with greater precaution, sweeping the Bees off very gently with the feather until there is not one left on it. This operation, I perform without gloves, or any other protection, armed only with my pipe; and, for ten times that I treat them after this fashion, I seldom receive one sting, even when I act unassisted.

The combs being all removed, the swarm remains as completely destitute of food as it was on the day of its emigration, and I replace it on its board in the same spot it occupied when full, and leave it till the evening, by which time the Bees will be clustered together like a new swarm. During the whole of the day, which I shall suppose to be fine, they occupy themselves with great earnestness cleansing their house, and making such a noise in removing the little fragments of wax that have fallen on the board, that any one who did not know it had been emptied, would take it for the best and strongest of the hives. Before night, when they are all quiet, I throw a few whiffs of smoke in at the door of the hive which I mean my deprived swarm to enter, and which should be its next neighbour, on the right hand or the left; then, turning it up and resting it on the ground, I sprinkle it all over with honey, especially between the combs where I perceive the greatest number of Bees: five or six table-spoonfuls generally suffice; at other times three or four times as many are required. If too little were given, the new comers might not be well received; there might be some fighting; and, by giving too much, we run the risk of drowning them.—One should cease the sprinkling when the Bees begin to climb up above the combs, and shelter themselves on the sides of the hive, this done, I replace the hive on its board, which should jut out about seven or eight inches, raising the hive up in the front with two little bits of stick, so as to leave a division of an inch between it and the boards to give free access to the Bees. I also spread a table cloth upon the ground before it, raising and fixing one end of it on the boards by means of two bits of stick, that are placed as a temporary support to the hive. I then take the hive that was deprived of its combs in the morning, and with one shake, throw the Bees out of it upon the table-cloth, which they instantly begin to ascend; while, by the help of a long wooden spoon, I guide them to the door of the one that is placed for their reception. A few spoonfuls of the Bees raised and laid down at the door of the hive will set the example, they enter at once, and the others follow quickly flapping their wings and sipping with delight the drops of honey that come in their way, or officiously licking and cleaning those first inhabitants that have received the sprinkling, and with whom they mingle and live henceforth on good terms, one division of the new comers always cluster on the front of the hive, which they enter during the night without disturbance, much pleased to join their companions.

Next morning, early, it is necessary to take away the table-cloth and the bits of stick that were placed to raise up the hive and facilitate the entrance of the bees, and for some days the door should be left open as wide as possible. The hive should also be moved a little to the right or left, that it may stand precisely in the centre of the place they both occupied before the union.

I have frequently united three swarms in the same manner, and with the same success, taking care only to empty in the morning those on each side, and to make the bees enter the middle one in the evening, after it has been sprinkled with honey. In this case I do not remove the one that unites the three swarms."

I have adhered strictly to these directions except in "raising and fixing the table-cloth to the board,"—making the bees ascend, I have always found to be a slow process, but placing the hive they are to join over them when heaped upon the cloth, is much quicker and equally successful.

Old stocks that are rendered weak by swarming, or by having too much honey taken from them, may be united in the same manner, with this difference only, that double the quantity of honey should be used in sprinkling.

If a stock of Bees containing fifteen or twenty pounds of honey in September, be carefully managed during the winter, which consists in narrowing the entrance to exclude robbers, carefully covering the hive with a milk-pan, and raising it from the board every month or six weeks to clean it, no doubt can be entertained to its affording a good box of honey.

CHAPTER IX.

Manner of feeding weak Stocks, and the time most appropriate for this operation.

AUTUMN and Spring are the most proper seasons for supplying weak stocks with food. Bees ought never to be fed during the winter, as food given at that time, not only causes disease, but induces them to go out of their hives, when many of them perish from cold.

Food should be administered only at night, and the sooner after sunset the better; the vessel in which it is given ought to be carefully removed by sunrise the next morning, or robbers will be attracted to the hive by the smell of the honey and far more injury be sustained from them, than the benefit arising to the Bees, from the food given. In feeding, therefore, it will be necessary to observe the greatest neatness. In Autumn, Bees should be fed copiously, those hives containing less than fifteen pounds of honey must be made up to that weight by feeding; the most effectual method I have been able to devise is to excavate a board of four or five inches in thickness, so as to allow a soup plate, or pewter dish to fit into it without rising above its level; this dish may be filled with honey, and covered with pieces of paper to prevent the Bees from being drowned, it may then be placed under the hive at sunset, and a napkin tied round the bottom of it, to prevent any of the Bees from making their escape; in this manner three or four pounds of honey may be given at one time, so that twice feeding, it is supposed will be sufficient for any hive, for if more than this quantity is wanted, the stock must be joined to another as directed in Chapter VIII. Should the honey be very thick, a small quantity of warm water may be added to it, in the proportion of half a pint to three pounds of honey, observing to mix them well together.

If the honey be much candied it maybe placed over a fire for a few minutes till it becomes liquid—another plan of feeding is to prepare a rim of straw, or a wooden hoop, the exact size of the hive, and four inches deep, within which place the dish of honey, and put the hive over it, making the union secure with a napkin.

In the Spring, Bees should be fed sparingly, three or four ounces of honey twice in the week, will be found amply sufficient; the easiest method of giving these small quantities is by a vessel of tin, upon the same principle as a bird's fountain, holding about a pound or a pound and half (see fig. 9.) the projecting trough or mouth, must be put in at the entrance of the hive, it is one inch and three quarters wide, and three inches and a half long, covered with a perforated tin: this vessel being filled with honey, has only to be placed in the hive at night, and removed in the morning, the feeder itself effectually stopping up the entrance of the hive.

Some persons feed their Bees at the top of the hive, but it is much too tedious and sparing a way, in my opinion for Autumn feeding, in the Spring it does very well. This feeder is of wood with a cover of glass, it has a hole through its centre, corresponding with one at the top of the hive, which enables the Bees to pass into it, and take the honey—I believe it was invented by Wildman.

CHAPTER X.

Food proper for weak Hives.

I AM decidedly of opinion that Bees fed in the Autumn should have honey, in preference to any other kind of food. Mr. Huish recommends "eight pounds of honey, six pounds of water, a bottle of white wine, and a pound of sugar, boiled and skimmed, to be bottled for use," he adds, "the most advisable method is not to make more than is immediately wanted, because there is some danger of its fermenting." Now if the Bees are allowed to store a quantity of this, or any other similar kind of food in their hives, will it not in all probability ferment *there* also? this is my reason for recommending honey only—indeed I have never seen Bees so healthy as those fed on the simple mixture of honey and water. In Spring, other kinds of food may answer very well, as a small portion only is given at a time, and very little of it deposited in the combs. A very good Spring food may be made with honey and sweet wort, or with raw sugar and sweet wort, boiled and skimmed.

The proportions would be, one pound of sugar, or half a pound of honey, to two pints of strong wort.

For feeding weak stocks many things have been prescribed, but nothing is so proper and natural as honey, but I dislike feeding altogether, except a little in the beginning of the year, through the lateness of the Spring some hives (otherwise sufficiently supplied) may require it. Early swarms may also require a little honey when the weather proves unfavourable for their collecting it the four or five first days succeeding their being hived, but in both these cases a very small quantity will be found sufficient. Autumn feeding very rarely answers the purpose of the proprietor. Uniting the weak stocks at that season as directed in Chapter VIII. will be found much more advantageous.

CHAPTER XI.

Enemies of Bees and means of overcoming them.

GELIEU says, "that nothing is more prejudicial to Bees than ignorant attention, their most formidable enemies are, perhaps, their possessors, who busy themselves to torment them, and weaken and kill them by too much care. In Winter, they hurt them by shutting them up, and in Spring, the giving them a little honey is not always attended to, neither is the guarding them from moths, which, at that time, make the greatest havoc, nor is the narrowing of the entrances to prevent them being robbed. Some people suffocate them in Autumn, that they may possess themselves of their provisions; and others take out the best of the honey, and often too much of it, and so expose them to die of hunger.

"I therefore place, in the foremost rank of their enemies, those of their possessors, who, by their own ignorance and inexperience, hinder them from prospering and multiplying."—To all this I am sorry to say that I can bear testimony.

Amongst the enemies of Bees are enumerated, ants, moths, birds, poultry, mice, wasps, and spiders. Ants perhaps are their least dangerous enemies, for though they cannot sting them, they carry them to a distance.

Ants may be destroyed by pouring boiling water into their nests, and the operation will be greatly assisted by making holes into them with a sharp stick, so as to allow the water to flow readily to the bottom of them. Mr. Huish says, "to preserve my Bees from these vermin, I always fasten a piece of sheep's skin, with its wool on round the bottom of the pedestal," it has been said, that these insects dislike both garlick and shalots and they will not harbour in the ground in which these vegetables are grown.

Moths[10] are by far their most formidable and dangerous enemies, great numbers of hives are destroyed by them every year, it therefore requires the utmost vigilance on the part of the Bee-keeper to defend his favourites from these most powerful assailants. It is in the caterpillar state that they commit their ravages, and it is truly astonishing to observe the rapidity with which they destroy a hive, when they get established in it. It must be observed that hives managed upon the Depriving System, that are expected to stand for ten, fifteen or even *twenty years* are much more subject to the incursions of moths, than those which are destroyed every year. The best method of preventing their increase is the frequent cleansing of the hive floors, for the female generally deposits her eggs between the hive and the board on which it stands, or in the dust that accumulates at the bottom.

Upon removing the hive the moths maybe seen in the *larva* state upon the floor, and are easily destroyed. Moths and spiders, says Dr. Bevan, "should be watched and destroyed, in an evening, as at that time the former are hovering about, and the latter laying their snares." He also recommends a frequent cleaning of the hive floors.—Huish says, "the butterfly of the moth that redoubtable enemy of the Bee, appears in April, and continues until October. Destroy them as much as possible; frighten not away the bats which fly about the hives, as they devour a great number of them." He says also, "I would always advise an Apiarian to fix his attention particularly on a hive, the Bees of which appear to be in inaction, whilst the Bees of other hives are in activity.—If this inaction continue for ten days, or a fortnight, not a moment then should be lost in examining the hive, and the ravages of the moth will soon present themselves."

[10] Entomology designates two species of wax moths; the greater is the most common *gallerea cereana*, and the smaller *gallerea alvearia*.

Amongst birds, that little marauder the blue titmouse, (*parus major of Linneus*) stands the foremost as their enemy, for, says Purchas, "she will eat ten or twelve Bees at a time, and by and by be ready for more; she feeds her young ones also with them. She eats not the whole Bee, but divides it in the middle, pulls out the ba*gge* of honey, l*i*tting drop the skinny outside, in the spring she watches for them in the willow and sallow trees, and takes them when they are at their work." Destroy their nests in breeding time, and shoot them in winter. Lapoutre, a French Naturalist, assures us, "that he saw under a tree in which there was a tom-tit's nest, a surprising quantity of the scaly parts of Bees, which this bird had dropped from its nest." Sparrows and swallows have both an ill name, but I could never observe any great hurt done by either of them. Poultry will occasionally destroy Bees, for I once recollect seeing a hen and her brood pay dearly for their freedom with a hive, the chickens were all stung to death, and the hen escaped only with her life.

From mice, the surest safeguard is the single pedestal (see fig. 1, 2 and 3).

The destruction of Queen Wasps in the Spring, and wasps nests in the Summer, will prove the best security against these formidable enemies: Queen Wasps are seen in April and May, and are very easily captured, every one which is then destroyed, would probably have been the founder of a nest, which may be computed at 30,000, at least.

In Autumn, it is very common for Bees to rob and plunder each other's hives the best remedy for this evil, as well as to guard against wasps, is to contract the entrances; to effect this object, I would recommend a copper guard to be attached to each hive; (see fig. 10, and page 18.) the wooden one of Espenasse, as well as Huish's tin guard, I have found very inconvenient, which induced me to construct the one referred to. It is made of thin copper and stitched to the hive with copper wire, it has two sliders, one pierced with a number of small holes, and the other having one only, but large enough to allow the passage of one Bee through it. The advantages of this, above all other guards, arises from its not projecting beyond the hive, the alighting board not being encumbered by it, and the angles formed by Huish's with the hive, so annoying to the Bees, being avoided.

"In a word," says Purchas, "if you desire to have your Bees thrive, and prosper, keep them well from winds and wet, heat and cold, *destroy their enemies*, and let them enjoy a sufficiency of food gotten by their own industry; and if there be a want in some, timely supply them, and doubt not if, by God's blessing on your endeavours, the increase and prosperity of your Bees."

CHAPTER XII.

Directions for Hiving Swarms.

BEES managed upon the Depriving System, rarely swarm and are seldom found clustering at the mouth of the hive, for every bright hour during the honey season they seem to turn to profit, when however Clustering or Swarming takes place, it generally arises from the Box not having been put on sufficiently early in the season, or for the want of a second Box; if a swarm should from these causes be compelled to leave the hive, let it be put into a new improved Cottage Hive, (see fig. 3.) in the usual manner.

I have always adopted the plan of placing my swarms where they are to remain, within ten or fifteen minutes after the time of their being hived, and in this practice Gelieu agrees with me, for he says, "most people who have Bees allow their Swarms to remain till the evening in the place where they have alighted, and do not move them to the Apiary till after sunset, this method has many inconveniences.

As soon as a swarm has congregated in the new hive, and seems to be at ease in it, the most industrious amongst the Bees fly off to the fields, but with a great many precautions. They descend the front of the hive, and turn to every side to examine it thoroughly, then take flight, and make some circles in the air in order to reconnoitre their new abode, they do the same in returning. If the Swarm has taken flight in the morning, the same Bees make several excursions during the day, and each time with less precaution, as becoming familiarized with their dwelling, they are less afraid of mistaking it, and thus, next morning, supposing themselves in the same place, they take wing without having observed where they have spent the night, and surprised at their return not to find the hive in the same place, they fly about all day in search of it, until they perish with fatigue and despair. Thus many hundreds of the most industrious labourers are lost, and this may be entirely avoided, if the Swarms be removed as soon as the Bees are perceived coming out—this sign alone is sufficient.

Sometimes I do not even wait till all the Bees clustered in front or on the sides of the Hive, are reunited to their companions in the interior, as they are never long in being so; and this plan has always fully succeeded with me."

Experience has long since proved, that the custom of beating warmingpans and the like, at the time a swarm leaves the hive is perfectly useless, as well as the ridiculous practice of dressing the Hive, as it is called, by drenching it with beer, honey, fennel, &c.; the former is considered by persons of observation, actually to prevent the Bees from alighting so soon as they would otherwise do; and the latter frequently to compel them to leave the hive. The best method is to watch the Swarm in silence, and after it has once collected, to lose no time in hiving it into a *new*, *clean* and *dry* Hive. Much time and trouble may be spared the Bees, if the loose straw be removed from the interior of the hive, the best method of effecting which, is to singe them off with a wax taper, and afterwards to remove them with a hard brush.

I have for many years past discontinued the use of sticks across the interior of my Hives, for they cause much unnecessary trouble to the Bees in the construction of their combs; every facility should be given to a fresh swarm in their labors, for they have much to do, as Dr. Aikin has very beautifully said for them in the

SONG OF THE BEES.

We watch for the light of the morn to break,

And colour the grey eastern sky

With its blended hues of saffron and lake,

Then we say to each other, "Awake, Awake!

For our winter's honey is all to make,

And our bread for a long supply."

Then off we hie to the hill and the dell,

To the field, the wild wood and bower;

In the columbine's horn we love to dwell;

To dip in the lily with snow-white bell,

To search the balm in its odorous cell,

The thyme and the rosemary flower.

We seek for the bloom of the eglantine,

The lime, painted thistle, and brier,

And follow the course of the wandering vine,

Whether it trail on the earth supine,

Or round the aspiring tree-top twine,

And reach for a stage still higher.

As each for the good of the whole is bent,

And stores up its treasures for all,

We hope for an evening with heart's content,

For the winter of life without lament

That summer is gone, with its hours mis-spent,

And the harvest is past recall.

And not only do sticks across the hive cause much unnecessary trouble to the Bees in the construction of their combs, but render their extraction almost impossible; for in this System it becomes necessary, after a Hive has stood seven or eight years, to cut out part of its combs, which by that time will have become very black, very thick and the cells, from the number of Bees hatched in them, (every one leaving a deposit) much contracted. The times best suited for this operation will be March and September; if performed in March, two leaves of comb may be taken, if in September, one only; it is a very simple process, and easily accomplished with the aid of a little tobacco smoke, and a knife (fig. 6.) which I will hereafter describe.

Gelieu says, in 1814, "I have several Stocks from twelve to twenty years old that are as prosperous as the young ones, and one stock *twenty-five years old.*"

I cannot say so much as this, but I do not in the least doubt the truth of it; fifteen years is the longest time that I have kept a stock, and the reason of my losing it at the expiration of that period was from the decay of the Hive, it being badly made and not painted; its annual profit was never less than forty, and some years, fifty shillings.

I would recommend every person who keeps Bees, to have a few well painted new hives always by him, that each hive be weighed, and its weight upon a ticket of lead fastened to it, the board also upon which the hive stands should be weighed.

CHAPTER XIII.

Description of a knife for cutting out the combs (fig. 6.)

THIS knife, which is so simple in its construction, and so easily used, deserves to be made generally known. Gelieu, to whom Apiarians are much indebted, tells us, that in Switzerland it is commonly used, and that the combs from hives of any shape or materials are extracted without any difficulty. It is formed of a slip of steel (see fig. 6.) two feet long, by an eighth of an inch thick, the handle is twenty inches long, by half an inch broad, the turn-down blade of two inches in length is spear pointed, sharp on the edges, and bent so as to form an angle of 90 degrees with the handle; the other blade is two inches long, by one and a half broad, and sharpened all round; the broad blade cuts and separates the combs from the sides of the Hive, and the spear point, which is also sharp on each side, admits, from its direction and narrowness, of being introduced between the combs to loosen them from the top of the hive.

CHAPTER XIV.

Remedies proposed, as cures, for the Stings of Bees.

"THE sooner the Sting is extracted," says Dr. Bevan, "the less venom is ejected, and consequently less inflammation induced. To alleviate the irritation, numberless remedies have been proposed, of the most opposite kind and uncertain effect; as oil, vinegar, bruised parsley, burnet, mallow, or the leaves of any succulent vegetable, (renewed as soon as warm, and probably therefore, operating by cold alone) honey, indigo dissolved in water, &c., &c., the most effectual remedy appears to be Aq. Ammon. or Spirits of Hartshorn, nor is this surprising, when we consider that the venom of the Bee is evidently Acid."

I have known both the pain and inflammation greatly relieved by Laudanum; but for myself I could never experience the slightest relief from its application.

Its effects are very different in different persons for whilst a single Sting will sometimes produce alarming symptoms in one person another may receive many without being inconvenienced either by pain or swelling; this I am sorry to observe is not my case, for a single sting causes me the most acute pain, accompanied by violent swelling and inflammation, which lasts two or three days; the above remedies have proved totally ineffectual in affording me the slightest relief, but I consider myself very fortunate in having lately met with almost a perfect cure, and it is as immediate as it is effectual; I have much pleasure in communicating it, for notwithstanding every precaution, persons who are much amongst Bees occasionally meet with a sting. The method I have of late adopted, by which the pain is instantly removed, and both the swelling and inflammation prevented, is to pull out the sting as soon as possible, and take a piece of iron and heat it in the fire, or for want of that, take a live coal, (if of wood the better, because it lasts longer) and hold it as near to the place as I can possibly endure it, for five minutes; if from this application a sensation of heart should be occasioned, a little oil of Turpentine or Goulard Cerate must be applied.

During the last three years I have used for myself and those about me, who might chance to meet with a sting, a still more effectual remedy than the above, and as its application is more simple it is certainly to be preferred. It consists in applying the least possible quantity of *Liquor potassæ* immediately upon removing the sting either with a fine camel's hair pencil, a sharp pen, or even with the point of a needle. The venom of the Bee being an acid, this very powerful alkali consequently neutralizes it, the pain is instantly removed and neither swelling nor inflammation follow. Should too large a quantity of this alkali be used, (as from the hurry in which it is usually sought after frequently happens) the part should be plunged into cold water, or a scar will be the consequence, which will last for some days. I have found the quicker the application, the more effectual the cure.

CHAPTER XV.

Means of preventing the Bees from being stolen, especially in the Country.

I ENTIRELY disapprove of Houses of every description for Bees as a protection against weather, because they tend in various ways most effectually to destroy the lives of those valuable insects, for, as before stated, they form a shelter for and promote the increase of some of their most determined enemies; but there are other things to provide against, not only robbers of their own species, but those who would carry away hives as, well as honey, and at the request of a friend who has twice been deprived of all his hives by this latter description of Robbers,—I am induced to give what in my opinion is the best kind of house to answer this purpose, and likely to be the least injurious to the Bees. It may be sufficiently wide to allow of six hives with a milk-pan upon each, to be placed in a line, which will occupy about eleven feet, its depth may be about twenty-two inches, the top and ends weather boarded, the front and back composed of wood or iron bars, placed so that a small Hive or box will not pass between them, the front ones may be fixed, but those at the back must be made to remove, and secured by a lock.

At fifteen inches from the ground must be placed two pieces of wood, about four inches square, well secured by cross pieces at the ends of the house and by an upright in the middle, upon these pieces place the boards on which the hives stand, sufficient height must be given to allow of two Boxes to be placed upon a hive, three feet at least should be allowed.

Another method of security may be adopted which is equally effective, and as the inconveniences of a House are avoided perhaps it is the most preferable. It is to have a stout iron ring of four inches in diameter, a little flattened at the sides so as to become oval, worked in the back of the Hive a little below its centre, when the Hives are placed in a line upon pedestals of equal height a chain may be passed through these rings and locked at each end.

CHAPTER XVI.

Method of Dislodging Bees from Trees or Buildings, and putting them into Hives.

IT very rarely happens that Bees in Buildings or in Trees survive the Winter, cold and humidity usually kill them, but the comb and the little honey remaining in it induces others to visit their abode in the Spring, and which in all probability will be followed by a Swarm from some neighbouring Apiary in May or June, and on this account Bees are said *always* to be there, a little observation will prove the incorrectness of this statement, and I believe the only criterion by which it can be proved that Bees have *actually existed in such places through the Winter*, is to see them carrying in pellets of farina upon their legs in March and the early part of April.

The most proper time for dislodging a Stock, which from having fixed upon some warm and dry abode, has lived through the Winter, is towards the end of May or the beginning of June, but a Swarm ought to be removed upon the day of its arrival, or as soon afterwards as possible.

The only method that I can recommend at all likely to succeed, and which I have found to answer the purpose, is to lay the combs quite bare, and then to cut them out one by one, sweeping the Bees with a feather or the back of the knife, from each piece of comb into an empty hive.

Should the Queen not be observed during this process, it will be advisable to take a leaf of Comb that is filled with eggs or brood, and fasten within side the hive intended for the Bees, that they may be enabled to make a new one, should their original Queen have been killed or lost during this operation; for it is now proved beyond all doubt that they have the power of doing this, provided they have either eggs or brood in their hive.

The use of a little tobacco smoke throughout this operation will be necessary, a cigar in the mouth of each operator (for there must be two persons, if not more, engaged in it) will be found sufficient.

The Queen Bee may easily be distinguished by a common observer, her wings are very short, not extending beyond one half of her body, which is much longer, and more pointed than that of the working Bees, her legs are copper coloured, and her body brown.

The combs being all removed and the Bees swept off them into the hive, it must now be put upon a board and placed exactly where the Bees entered before they were disturbed, and, if possible, it should remain in this situation till Autumn, but its remaining for a week or two is absolutely necessary; should there be many combs filled with brood, three or four of them may be placed in the hive, by putting some wooden pegs at the top of it to keep them at proper distances, and allowing them to rest upon the hive-board.

This, however, is a tedious operation and very seldom pays for the trouble it occasions; for stocks thus obtained are generally weak and require feeding, or to be united to others to keep them alive through the winter. It is therefore more to the advantage of the proprietor to cause the combs in trees or buildings, in which Bees have died, to be destroyed, and the places effectually stopped up with clay or mortar.

CHAPTER XVII.

Description of Nutt's newly invented Hive, for obtaining Honey without destroying the Bees.

THIS Hive consists of three collateral Boxes, (see fig. 8.) the centre one fourteen inches, and the side ones ten and a half square inside measure, depth eleven inches; the middle box has a number of holes in the top three quarters of an inch in diameter, bored in the circumference of a circle six inches in diameter, with one in its centre, over which a glass is placed. The side boxes (*aa*) have each a hole of four inches square in the centre of the top, into which is fitted a piece of tin pierced with small holes, and in its centre a hole of an inch in diameter, wherein is placed a tube of tin of the same size, reaching nearly to the bottom of the Box, and supported by a shoulder resting upon the square tin, which is also perforated. This square tin and cylinder constitute the ventilator, the opening in the Boxes is covered by a piece of wood, (*bb*) the Boxes have each a glazed window (*c*). The centre Box communicates with the side ones by a grating cut in the wood and corresponding with each other, this communication is cut off when necessary by sliding tins.

The method that I have adopted for protecting this Hive from wet, is to cover the glass with a common straw hive, upon which a milk-pan is placed, and each of the side boxes by two pieces of board eleven inches by fifteen, put together so as to form an angle of ninety degrees, resembling the roof of a cottage. The whole should be made of inch deal and well painted, the board upon which the three boxes stand must be of two inch deal and of one piece, except the alighting board, which is nailed on, three strong pieces of two inch deal will be necessary on the under side of the board as braces to prevent its warping.

In this I have differed a little from Mr. Nutt, because I think the solid board more substantial and less likely to harbour insects, his being hollow to allow the bees to escape at the time of deprivation, when the communication between the centre and the side boxes is cut off by means of the sliding tins, for which I have made provision by having an entrance at the back of each of the side boxes, to which is attached a copper slider, (fig. 10.) but without the entrance for a single Bee.

The middle Box must be stocked in the usual way, by hiving a swarm into it, and if the swarm be not a large one, it will be better to put two swarms into it as directed in Chapter VIII. It will be necessary to have both the top and side tins securely fastened to the centre box at the time of hiving, and till after the union, when they may be removed, thereby giving the bees the full range of all the boxes as well as of the glass upon the centre one, nothing further will be required till the next Summer, except carefully placing the roofs to defend the boxes from wet.—"Perhaps," says Mr. Taylor, "there is nothing more prejudicial to Bees than the moisture they engender during the Winter season, particularly after frost, and in certain states of the atmosphere. It accumulates on the top and sides of the pavilion, moulding and rendering offensive the combs, and producing disease. For this reason boxes with flat roofs have been objected to. I have tried different experiments to obviate this serious evil, and have found nothing better than the practice of condensing the vapour as much as possible and conveying it away. This I have done for several years by means of the bell-glasses.

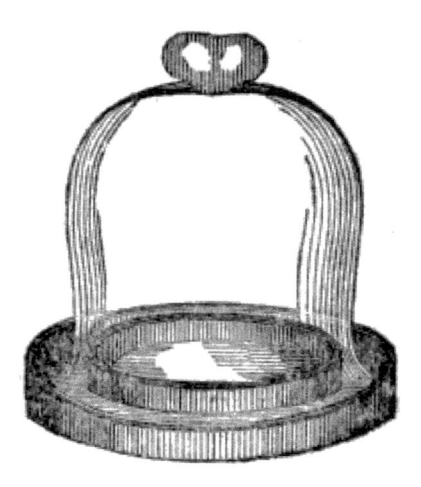

At the beginning of Winter I place over the holes on the top of the pavilion, pieces of perforated zinc, and on these I set the glasses, each within a circular leaden or zinc trough, open in the centre. As the exhalation rises from the warmth below, it is condensed on the glass, and received, often in considerable quantity, in the troughs. An imperceptible current of air is thus produced, of great advantage to the inmates; for ventilation is as much wanted in winter as in summer, and particularly when the population is numerous. The holes at the top of the glasses may be left open to assist this, for of two evils it is better to have too much than too little air. Nor, with good protection from weather, need the effects of cold be apprehended; for the Bees, (if not weak in numbers) will always of themselves generate sufficient warmth, and a dry, cold season is better withstood than a mild, moist one, particularly after a good honey year."

This leaden or zinc trough of Mr. Taylor's exactly resembles in form the feeder mentioned in page 56.

In the following Summer, when the Bees are working in the side boxes and the glass, the ventilation must be particularly attended to, (for upon this *alone* depends the success of the hive,) and here I would especially recommend Mr. Taylor's ventilator as given in page 25; a small thermometer should be kept in one of the ventilating tubes, and when it is observed above ninety degrees of Fahrenheit, the covers (*bb*) must be taken off, and should the temperature of the boxes be found at, or above a hundred, the perforated copper slides at the back of the side boxes, must be used, for if it reaches to a hundred-and-thirty, a Swarm in all probability will leave the hive, which next to starvation is most fatal to this mode of treatment. When either of the side boxes or the glass upon the top of the centre one is filled with honey and sealed up, introduce the dividing tin; if the glass, remove it immediately thirty or forty yards from the hive without changing its position, leaving room at the bottom for the Bees to escape, which they will very soon do, but if either of the side boxes are to be taken away, open the copper slider at the back of the box, and in less than an hour from the time of the dividing tin being introduced, the Bees will have made their escape and joined the centre box, it may then be removed, emptied and replaced, or another may be substituted for it. All operations must be performed upon a fine and bright day. The entrance to the centre box should be opened to its fullest extent, by removing both the sliders from the first of April till the first of September.

CHAPTER XVIII.

The Apiarian's Monthly Manual, or Hints for the Management of Bees for every month in the year, upon the Depriving System.

JANUARY.

Should the cold be intense, no operation whatever should be performed on the Bees that requires the removal of the hives. If snow be on the ground keep the perforated sliders (page 18.) closely down that air may be admitted, but not a Bee allowed to escape until it be thawed; but immediately upon the disappearance of the snow remove the slider, and give them full liberty. I have known many stocks lost by not attending to this precaution, and more especially after a long confinement, do not suffer the snow to melt either upon the covers or hive-boards, but brush it off every day as it falls. Attend regularly to the condenser (page 90.) which to boxes with flat roofs is a very necessary and useful appendage.

FEBRUARY.

Upon a mild day in this month let the floor board of each hive be cleaned, and a little food administered, should the stock of honey be very low. See that the coverings be sound, and that no moisture comes upon the top of the hives. Should it be found that any of the hives have perished, which will sometimes occur, and from causes which cannot be exactly ascertained, let them be immediately removed, and the honey which they contain taken out, and reserved for feeding those that may require it.

MARCH.

Clean the hive-boards again, and should any of the stocks require feeding, supply them, attending strictly to the directions given in Chap. IX. Towards the end of this month place a vessel, containing water, near the Bees, as directed in page 3. This also will be found a good time to examine the pedestals upon which the hives stand, for after remaining some years in the ground they are subject to decay at a few inches below its surface, especially if regard was not paid to the quality of the timber at the time of fixing them.

APRIL.

Clean the hive-boards for the last time, and supply food, if required, as before directed. The Wax-moth, that redoubtable enemy to Bees, appears this month; they may be seen frequently at twilight running upon the outside of the hives: destroy them as much as possible, and, as Huish says, "frighten not away the Bats that fly about the hives, for they destroy numbers of them." A full supply of small hives, boxes, glasses and adapters should now be provided, old ones cleaned, or new ones purchased. A few large hives also should be ready, for if from inattention to giving room and ventilation, a swarm should be compelled to leave their hive, they will be wanted.

Weak hives are now very subject to an attack from robbers, the best protection that can be afforded them is the slider page 18, with the help of which three or four Bees will guard the entrance more effectually than many times that number without it.

MAY.

The time will now have arrived for supplying each stock with a small hive or other receptacle for honey, as directed in Chap. 4, and should the season be a favorable one, the supply even of a second may be found necessary before the end of the month. Continue to destroy Queen wasps and hornets, and to watch carefully for moths. Should the bees of any hive appear inactive at this time, or should they not be seen to carry in pellets of farina whilst others are doing it, and this inaction continue for eight or ten days, lose no time in examining the hive, and should the moths have begun their work of destruction, which may be known by seeing the combs joined together by their silken webs, cut away the combs affected with a sharp knife, and the hive may, perhaps, be saved.

JUNE.

Strict attention should now be paid to *room* and *ventilation*, for, as has been said in page 24, if both these be carefully observed, *swarming may be prevented altogether*. Swarms may now be purchased as directed in Chap. II. About the middle of the month, in good seasons, small hives and glasses may be taken off, full directions for which may be found in Chap. V.

At the end of the month look for wasps'-nests and destroy them; a very easy and effectual method of doing it is to fill a common squib or serpent case with a mixture of sulphur and gun-powder, in equal parts, with a very small quantity of nitre all finely powdered and rammed very hard into the case, set fire to it by means of touch-paper, and when in a state of ignition, stick it into the hole of the nest and place your foot upon it, when it ceases to burn let a person with a spade turn out the nest; in this manner a great number may be effectually destroyed in one night.

JULY.

Small hives and glasses must now be taken off as they are filled and sealed up, (and stored in cool places, observing to keep them in the same position as when standing upon the stocks,) and their places supplied by empty ones. Go on destroying wasps'-nests.

AUGUST.

Continue to take off hives and glasses as they are filled, but supply no fresh ones, the honey season being now chiefly over.

SEPTEMBER.

Small hives remaining upon the stocks that are only partially filled with honey may now be taken off, providing the stock will not be too much impoverished thereby; wherever the chance of this presents itself leave them on through the winter, or until they are emptied by the bees; those partially filled hives taken from rich stocks may be given to weak ones, now, or in the spring as required. Robbers will at this time be carrying on their depredations, and should a serious attack be observed the sliders must be used as before directed.

OCTOBER.

Examine the coverings to the hives that they be all sound, and that no rain be admitted through them; the entrances may now be narrowed, if Taylor's slider be used (page 18) the side with three openings will be most proper for this season.

NOVEMBER.

Clean the floor-boards of the hives, and see that they stand firmly on their pedestals, contract the entrance so that only one bee can come out at a time, for at this season mice are likely to lodge themselves in the hives.

DECEMBER.

The same attentions are necessary this month as in the two preceding, but if the cold should be intense the hives must not be removed.

FINIS.